Ex-Nilo: How God Brought Something Out of Nothing

By RoseAnn Salanitri

Acknowledgments

It would be a gross understatement for any author to acknowledge only a few people that were instrumental in inspiring and/or influencing their work. Our lifetime experiences and all of our relationships contribute to who we are as people and what we believe. Nature vs. nurture is not a legitimate choice. Both play major roles in our lives, and I am no exception.

That being said, this particular piece of work would not have been possible without the scholarly input of Pam Elder and her knowledge of Hebrew. I owe her a profound debt of gratitude, as does anyone else that benefits from this work. And I am also eternally grateful to my father for inspiring me to love both God and science.

But deserving of more accolades than Pam or my father is our Lord and Savior who, after years of persistent prayer, blessed me with the understanding of how He brought something out of nothing. The Holy Scriptures tell us to love the Lord our God with our minds as well as our entire beings. For me it would be impossible to do anything else. I am eternally grateful for the gift of knowing God and having been given a glimpse into the genesis of His creation.

Table of Contents

	Page
Introduction	i
Preface	iii
Chapter One:	
Asking the Right Questions	1
Chapter Two:	
Genesis 1:1 - Genesis 1:2	13
Chapter Three:	
Genesis 1:2	25
Chapter Four:	
Genesis 1:3	41

Introduction

Although the idea of Evolution began with the Greeks, Charles Darwin popularized it worldwide. His scientific approach for a subject that had remained the domain of religious sects launched a new philosophy that captured the minds of many. No longer would the vast majority of the educated domain attribute the observable world to the creation of an intelligent being; instead the search for natural explanations as opposed to supernatural explanations for the observable world became the scientific standard - and today, dominates as the only acceptable argument.

However, after over a century and a half of scientific research for Evolution and the origin of life, not only has Evolution not been scientifically proven, but significant questions remain unanswered. While some may argue that the absence of evidence is not evidence of absence, the absence of evidence plagues what has become the only scientific standard accepted throughout academia. And this standard is ferociously defended through censorship, intimidation, and the firing of those expressing a different point of view. (See "The Slaughter of the Dissidents" by Dr. Jerry Bergman.)

Censorship of ideas that contradict reigning and popularly held dogmas go back centuries. Ironically, Galileo, whose scientific theories were censored by a controlling religious system, was the most well-known victim of scientific persecution. Today, the tables have turned and it is the scientific community that victimizes the voices of opposition. Regardless of who is censoring whom, censorship is never a good thing. As a result, science has suffered the most. For the honest search for scientific truth must encourage even the most radical ideas.

This work will present perhaps some of the most radical ideas proposed in decades, and only those passionate enough to pursue truth regardless of their worldview will be brave enough to read it with an open mind.

Preface

Although the Creation account has been dismissed as fantasy by many (as is the Creator), a study of the Hebrew words used in Genesis 1:1-8 coupled with modern discoveries in physics have unveiled a startling explanation of the most provocative question in science today. For the Evolutionist that question would be: how did something naturally evolve from nothing; and for the Creationist it would be: how did God create something out of nothing?

Since Darwin first wrote "The Origin of Species"[i], the hypothesis of Evolution has had an impact on the educated world. Through intimidation and by strongly suggesting that the biblical account of Creation is nothing more than a crutch for the weak-minded, Evolution has indirectly served as the subliminal evangelist for atheism and agnosticism. It has also served as a tool for the foundation of the Progressive movement in the United States, with Woodrow Wilson acting as its founding father. Wilson applied his Darwinian beliefs to government, advocating that the Constitution as ratified over a century prior, was antiquated and should therefore be considered a living document in order for it to "evolve" with the human race.[ii] When this Darwinian application of scientific and political philosophies gained popularity, it provided a scientific explanation to support the many sneers from the intellectual elites that clearly intimidated Creationists with remarks like: "Surely, you can't really still believe in that stuff!"

In the new millennia the tides may be turning on the cynicism of Evolutionists, with the irony being that the emerging physics support the Creation account in ways never realized before. For those who are already balking at the notion that the biblical account of Creation is scientifically plausible - and yes, even supportable, the challenge will be to read this work with an open mind. You must consider that if God created everything as His Word tells us He did, then surely He must have created the physical laws to support His creation. Therefore, it is logical to look to the physical laws and phenomena for evidence of the veracity of His Word.

The key to understanding the physics that support how God brought something out of nothing lies in the familiar verses in Genesis' Creation account that are often glossed over by theologians and laypersons alike. They fail simply because they do not delve into the in-depth meaning of the words written by Moses[iii] approximately 3800 years ago - words that probably didn't make sense to the obedient Moses, and words that don't seem to make sense to the reader of the Scriptures today. However, coupling the deeper meaning of the Hebrew words with modern physics reveals the sequence of creation that simply could not have happened any other way.

For example, closer examination of the words for "formless," "void," and "water" when used in conjunction with the word "hovering" reaffirm recent discoveries in quantum mechanics, vacuum physics, and plasma physics by reputable and well-renown research laboratories. The creation of something being brought out of nothing can now be understood when studying Genesis and applying modern physics. All of this was before our very eyes for close to four millennia but we didn't have the science to understand what God was telling us. Now we do. And it is this understanding that leads us straight back to God as the Creator, as defined in the creation account.

Close-minded skeptics may still find reasons

not to believe but the rational and logical thinker should come to the knowledge of the truth: God is God; He is who He says He is; and He is the Creator.

As astounding as it may sound, the Bible supplies the blueprint and all of the answers to many provocative scientific questions regarding origins. It asserts, and now we can confirm, that the universe was supernaturally created by a being outside of the time/space continuum who told us exactly how He did it

The first chapter will contain a general explanation of where this work is leading. Each subsequent chapter will proceed with a more detailed analysis of the Hebrew words and how they accurately explain what happened scientifically on that first day of creation. The chapters will proceed with the sequence of creation as revealed on the first day of creation in the biblical text and then expound upon the many awesome revelations of the mysteries contained in those few verses regarding the first day -

revelations that existed four millennia before the scientific discoveries that confirm them came into existence!

Chapter One

Asking the Right Questions

Since Einstein unsuccessfully spent the last 30 years of his life looking for it, Grand Unification Theory (GUTs) has remained the golden fleece of physics. Einstein never found the answers he was looking for but he did ask the right questions. GUTs requires that the small and large forces of physics must not conflict[iv]. As significant as the discovery of this may be, it will only lead to another question: *how did something come out of nothing*?

Modern day physics has revealed that chaotic bursts of energy existed primevally in a Zero Point Energy Vacuum (ZPE), even in the absence of heat[v]. These chaotic bursts of energy are referred to as quantum fluctuations.[vi] In his famous "Feynman Diagram"[vii], Richard P. Feynman expanded upon this in his quantum electrodynamics theory, which gave rise to the concept that two particles can communicate and exchange information through the creation of a virtual particle[viii]. The developing sciences of vacuum physics, plasma physics, wave physics, and quantum electrodynamics provide the basis for understanding what existed at the time of creation, including the communication of information. This would be necessary for the continued creation of matter in any kind of an organized or repeatable fashion. Without the communication of information, the organization of quantum forces would be sporadic and incapable of forming the repeated and systematically spaced organization we see in the elements, as documented in the Periodic Table, as well as in the laws of physics and the codes that exist in the genome.

If God is who He says He is, it should be expected that these sciences will support the Creation account as set forth in Genesis, since information had to be created in the past to support what we observe today. Before the communication of information could happen, information must exist, and in Feynman's theory, it would provide the cause before the effect - although Feynman never touched upon it. Without the communication of information, systematically organized elements, as well as flora and fauna, and the reproduction within the animal kingdom that relies on codes would not exist. Therefore, the Evolutionist is faced with another profound question: how did information evolve without an intelligent creator?

Understanding that energy is contained within matter is not a foreign concept. Einstein understood and wrote about it in his famous equation E=MC^{2}[ix]. However, the transformation of energy into matter requires the understanding of the state wherein energy preceded matter in a ZPE environment[x]. Next we need to understand the catalyst that organized this primeval chaos into kinetic energy, then plasma, next into acoustical/radiation waves, and then into the cosmos - all with the capability of communicating and storing information. Again, the question reverts back to: what or who provided this all important catalyst, as well as who created the information? While modern day physics lays the groundwork for this understanding from a mechanical point of view, it is the Creation account that actually gives us the only plausible sequence while also defining - or identifying - the catalyst.

This sequence begins within the realm of quantum fluctuations within a vacuum[xi] and brings us to the logical conclusion that chaotic bursts of energy preceded matter[xii]. Although science confirms this, it must be more intricately understood. Once that is understood, then we can understand how energy is organized into waves and then into matter - or how something is brought out of nothing.

The physics in Genesis and the meaning of the full breath of the Hebrew words in the sequence asserts how plasma (scientifically referred to as the fourth state of water) is produced and how waves and radiation result from God's "hovering" over the void. The explanation given in Genesis is supported by experimentation and discoveries regarding quark-gluon plasma - most recently at ALICE (A Large Ion Collider Experiment) at Cern[xiii].

In the case of light (radiation waves)[xiv], there is an interaction between waves and particles forming what is commonly referred to as a wave/particle[xv] because light behaves as both. It is the one observable entity where matter in the form of a particle exists in a suspended state of a multi-dimensional wave that can far exceed the 11 dimensions predicted in string physics. Harmonic resonance that exists within waves during phase transitions are most likely responsible for the multi-dimensional universe that consists of more dimensions than we have the technology to count, since every phase within an acoustical/harmonic wave is capable of creating another dimension. Additionally, the application of pressure increases this phenomena exponentially if the pressure is incalculable. And...pressure of this kind upon the primordial void would have to have been applied from outside of the void. In other words, if the void and its expanded universe is considered to be the "natural" then the entity applying the pressure must be considered to be "supernatural" = or above the natural. This will be discussed in more detail in the next *phase* of this writing.

More important than understanding the dimensions that are capable of existing within a wave, is the understanding that a wave is the transitional link between nothing and something, as well as the transitional link between primordial chaos, organization, and information. And...it is the first thing God *spoke* into existence in Genesis, since audible speech requires waves, but it is not the first thing God created. To those that study Genesis 1:2-3, that might seem like an anti-scriptural statement, but it is a statement that is supported both by Scripture and physics.

Whether or not you accept that God is the Creator that spoke light into existence or that physics acknowledges that light had to be the vehicle that expanded the universe (as Dr. Alan Guth argues in *The Inflationary Universe*), it is the transitional mechanism that allows the merging of energy and particles through waves. However, as previously stated, light is not the first thing God created. Actually, God first introduced kinetic energy when He hovered over the formless void, which resulted in plasma - or the fourth state of water. The result of vibrating a vacuum and creating plasma has recently been discovered at CERN. Again, we must remember that Genesis clearly states in Genesis 1:2b, "And the Spirit of

God was hovering over the waters." The plasma that was created when God added kinetic energy to the void was then capable of propagating acoustical waves - or speech. Then God spoke.

He also provided the mechanism for the creation of acoustical waves that produced light. These waves also possessed the capability of recording and transmitting information. Furthermore, plasma physicists believe that plasma (the fourth state of water) exists throughout the entire cosmos[xvi], which is exactly what we are told in the Creation account.

Unbelievers are faced with discovering the mechanism and the catalyst for the transition of quantum fluctuations into plasma and waves, as they are faced with finding the catalyst that began the expansion of the universe, frequently called "The Big Bang."

On the other hand, Creationists have the answers to these questions in the Creation account. Extraordinarily, the answers to these questions were documented by Moses approximately 3800 years ago, well before the questions were asked and well before modern physics provided enough information for us to even begin asking the right questions, never mind understanding the answers.

For both believer and unbeliever, the search for these answers brings us back to the primordial realm where the energy in a vacuum first transforms into plasma and waves and then develops light and records information in the process. As a Creationist, my standard is Genesis, Chapter One, and this work is based on that standard. The wealth of the physics contained in Genesis both startled and amazed me. Additionally, I will rely on Proverbs 30:5 that states: "Every word of God is tested" (NASB). I have made every effort to appropriately analyze the Hebrew meanings of the words in the verses discussed. They have revealed that the secrets to understanding the creation of the universe - not only from a spiritual point of view but even more profoundly from an impartial scientific point of view - have always been there. It simply couldn't have happened any other way.

The assertion that all these understandings are provided for in the Creation account will come as a surprise to many, including those who have read Genesis ad nauseum. There are three reasons for this:

1. Those verses have been read without believing that they were merely symbolic in nature and that they did not hold the answers to the most provocative scientific questions imaginable;

2. We simply did not have the science to understand what God was telling us; and

3. Studying modern developments in physics as well as the original Hebrew words for the full meaning of the verses was necessary and the translations we have usually relied upon are grossly inadequate in order to successfully apply the physics.

Amazingly, Genesis holds the scientific key to understanding how God created something out of nothing. Understanding Ex-Nilo required an intelligent force that existed outside of the space/time dimension we live within - a force that is and was capable of organizing energy into matter, as well as being capable of creating information, as well as the mechanism for the recording, reproduction and the transmission of that information.

While Evolutionists may argue that energy transforming into matter does not require an intelligent force, they would be hard pressed to argue that the creation of information does not require an intelligent force or how life evolved from inanimate matter. When God hovered and then spoke, He introduced kinetic energy into the void that transformed the quantum fluctuations in the void into the plasma necessary for the mechanical formation of waves that enabled the recording and the transmission of information, as well as the wave vehicle that enabled the formation of radiation/light, charges, water, and the production of matter.

In this work, I will analyze the sequence of Creation beginning with Genesis 1:1, and aided with more in-depth Hebrew interpretations than most Bible translations include. In so doing, I will apply the physical laws that substantiate the Creation account and argue *how God brought something out of nothing.*

This work will be following the logical sequence of events documented in Genesis, which begins with a topic sentence. The topic sentence is followed by a description of the state of existence prior to the creation and then proceeds to the sequence God employed as He created time, space, and matter.

Chapter Two

Genesis 1:1- Genesis 1:2

"In the beginning, God created the heavens and the earth."

Many theologians interpret Genesis 1:1 in many ways, and progressive creationists as well as old age creationists create still more ways. Grammatically, both the Hebrew style of writing that starts with a statement and attaches a string of additional statements to it, and the English styles of journalism that starts with a topic sentence would support the simple reading of this verse as being a sentence that explains what the rest of the writing will be about.

As clear as this verse may be from a grammatical point of view, it also suggests profound scientific principles. These were first understood by Albert Einstein. Einstein understood that time cannot exist without space. He called this phenomenon the "space/time continuum."[xvii] Genesis 1:1, while grammatically establishing what will follow, also gives us the basis for the creation of time (the beginning), as well as matter (the earth), and the cosmos (the heavens). The Hebrew tells us that this is the overview of the events in creation. Some may argue that this simple sentence is not to be interpreted scientifically; however, as this work progresses, the sheer volume of instances such as these will make it harder to sustain that argument.

Genesis 1:2

"The earth was without form, and void;

and darkness was on the face of the deep. And the Spirit of God was hovering over the face of the waters."

There are several significant scientific things being established in this verse. I will break it down piece by piece, first with an explanation of the Hebrew meaning of the word and then the wealth of information brought to the understanding of what happened relying on the application of modern physics. In other words, first what the Hebrew says, and second, the science that supports it.

The earth

The first word "earth" is interesting. If we assign the word the connotative meaning without researching the full breath of the Hebrew meaning, we would think of solid land composed of dirt and a variety of other materials, which has made this particular word an enigma to Creationists. Although the Hebrew word used in this verse can also include the usual meanings for the word "earth" that we would expect when used elsewhere, there is also another meaning. It is simply a designation that sets the earth apart from the heavens or the sky.[xviii] In other words, another territory, environment or entity being described by God that separates earth as a region or a field that is different than what will become the heavens or the sky. As NASA explores the cosmos looking for signs of life and environments similar to what we have here on earth, we are continually discovering how unique an environment the earth provides - even when compared to other planets.

The word used for heavens confirms this, since it is a plural designation - meaning that there were two heavens - as distinguished from the one earth[xix]. This duality can be interpreted to mean the earth's atmosphere and what we call the cosmos. It can also be interpreted to mean the universe we observe or another dimension that exists outside of ours. It is logical to think this refers to the earth's atmosphere and the cosmos because it is difficult to infer from the passage that we are being told about the creation of another dimension. It is not a stretch at all to say that the

Scriptures speak of two distinct heavens as compared to one earth. Right now it is important to understand that there is a distinction made between the earth and the heavens.

Part Two, not included in this work, will discuss the separation of the waters from the waters that occurred on Day Two. To confirm that the earth is a separate entity, we will learn that there is a distinct difference between what will become the earth as we know it and the cosmos. It is also not a stretch to say that Moses was obedient when he recorded these words, as surely it must have seemed odd to him that God referred to the atmosphere and the cosmos as "the waters." From Moses position in time and space, observation appeared to contradict what he wrote. Moses could see that the atmosphere was composed of air and the cosmos was empty, except for the celestial bodies. From Moses' perspective, neither the atmosphere nor the cosmos appeared to be water, and yet that is what he wrote under inspiration and with great obedience. It isn't until modern day physics are applied that the Scripture's use of the word "waters" in verses 2, 6 and 7 is confirmed and makes sense. As a prelude to the writings that deal with Day Two, it should also be noted that Day Two sets the basis for what will be labeled as the "Big Bang Theory" in modern physics.

Without form and void

Genesis goes on to use the words "without form" and "void" to describe the qualities of this designated area. The words "without form" do not describe a mass of a continually morphing or undistinguishable structure lacking any physical shape, as one would imagine a bubble that is floating through the air, changing shape as it glides along. The Hebrew word for without form is tohuw. According to Strong's Comprehensive Concordance (hereinafter referred to as "Strong's") this word also denotes a primeval state of confusion and emptiness.

The Hebrew word for "void" in Strong's is bohuw, which simply means empty or a vacuity.

Scientifically, this equates to what we describe as a vacuum (a place entirely void of matter)[xx]. Again, Moses must have been puzzled by the oddity of using these two words in harmony. What is being said is that there was an empty, formless vacuum that contained confusion (chaos) that was in constant movement - although matter did not exist within it. Before experiments were made in vacuum physics, saying that a vacuum that contained nothing also contained chaos[xxi] that was in constant movement can be more than the unenlightened mind could have conceived. However, that is exactly what vacuum physics has discovered and it is exactly what this particular Scripture says.

When we try to understand such things today we have the benefit of vacuum experiments, which Moses wasn't privy to 3800 years ago. These experiments have revealed that true voids (as would have existed primordially) do not exist as we previously thought but contain unstable amounts of high-energy fields[xxii] known as quantum fluctuations. Apparently even "nothing" contains something: chaotic bursts of energy or to put in the Hebrew terms: an empty state of confusion and chaos that is in constant movement. Many understand that atoms are composed of energy charges that form the nucleus of the atom and the electrons within their shell configurations, but we don't understand how quantum fluctuations turned into kinetic energy; which turned into acoustical and radiation waves; and which ultimately turned into the elements we observe today. The Creation Scriptures in Genesis actually help us to understand exactly how this happens, which is supported by recent discoveries in both vacuum physics and plasma physics.[xxiii]

In verse 2, God is telling us that what would become the earth started as a formless void with unique and unexpected characteristics. Once again, modern day physics helps us to understand this better. Conversely, modern day physics would also reach a higher level of understanding if it acknowledged the Creation sequence, enhanced by the Hebrew translation of the words.

To recap: physics tells us that ZPEs (Zero Point Vacuums - or the purest form of vacuums devoid of matter) contain quantum fluctuations. Moses tells us the same thing in the Creation account but he goes further in stating that the chaos was also in a constant state of movement. Here we have the merging of Moses' Hebrew with modern day physics, and together they help us to understand what existed prior to the beginning of time - or more properly stated: prior to the beginning of the creation of the space/time continuum.

When charted out for comparison purposes on the following pages, the argument can be easily recognized.

Genesis Statement	Hebrew Word & Meaning	Physics Application
"In the beginning… the heavens…"	Re'shiyth: Strong's #7225 translates to time; Shamayim: Strong's #8064 translates into two domains: 1. the cosmos 2. the atmosphere.	Einstein's Theory of Relativity argues that time and space are interwoven, which means they have to have existed together, for when you create space, you create time. This is referred to as the space/time continuum.
"…the earth."	Erets Strong's #776 An environment or region that is distinguishably different from the cosmos	Although NASA has been looking for years, it has yet to find a planet with the unique qualities of the earth. It should also be noted that all planets are infinitely different from other celestial bodies,

		including the structure of stars.
"…without form and void."	Tohuw: Strong's #8414 Primordial confusion, and empty place without form Bo-hoo: a vacuity - or a vacuum, an empty place, a void that is in constant movement.	At the beginning of time there was nothing but a primordial vacuum with quantum fluctuations, which contained nothing but chaotic bursts of energy.

While modern day physics doesn't tell us what existed before the Big Bang, it does assume that there was some kind of vacuum with a singularity - a small dense piece of matter of some kind. Although NASA diligently searched for monopoles (believing they would hold the answer since their single polarity would allow for a concentration of the anticipated matter) - monopoles were never found. Furthermore, exactly what that piece of matter was has never been discovered and has evaded further explanation by the finest minds in science - that is until plasma physics suggested that the matter must have been a form of plasma.

The Bible also tells us that a void (vacuum = nothing) existed at the beginning (before time and space were created) that contained continuously moving chaos, and that there was a distinctive entity that was different from the rest of creation that God called "earth."[xxiv]

Physics also tells us that vacuums contain chaotic bursts of energy - or nothing with confusion in the form of quantum fluctuations. The Bible tells us that as well and goes a step further by stating that this was the condition before God intervened and created the universe we observe today. As we proceed, we will also learn that the Hebrew words tell us that this primordial state was in constant movement. And we will learn that this overlooked aspect of the Genesis verse is extremely important scientifically.

Chapter Three

Genesis 1:2: "and darkness was on the face of the deep..."

This verse is speaking about the primordial state before God intervened, or if you are an Evolutionist, this is the state that would have existed prior to the Big Bang. It should be noted that the Big Bang, which really is a misnomer since the theory is about expansion and not explosion, occurs on Day Two and will be discussed subsequently in another work - God willing.

At this point, time and space did not exist - and neither did light. The word "face" is used to refer to the edge of the void. This verse is telling us that darkness surrounded the void, which means that the quantum fluctuations that exist within a Zero Point Energy system - or a vacuum - did not extend beyond the vacuum or the void. It also tells us that the void had an edge or was a confined environment. Once again, there is total harmony with the Creation account and the Big Bang Theory that stipulates that all the universe once existed in an extremely compact state before the expansion.

This assertion about the pre-creation universe is also significant from a physics standpoint when we consider light. We can easily understand that light requires a wave in order for it to be a wave/particle. At this point, water didn't exist yet, God hadn't spoken, and acoustical (sound) and radiation waves (light) had not been created. In this sense, the biblical account goes beyond the Big Bang Theory in that it is describing a time prior to the existence of matter, which the Big Bang assumes existed in the form of a singularity - or an incomprehensively small and dense piece of matter through which the entire universe evolved. Of course this raises many questions for Big Bang enthusiasts, like what was this form of matter, how did it come into being, and what caused it to expand. These questions do not exist in the Creation account, which this work is now extolling as being scientifically more plausible than any scientific explanation proffered to date.

Furthermore, the meaning of the word "deep" supports the science we have come to understand. In Hebrew it means something that is in a continuous state of movement. The importance of this being in a continuous state of movement will be developed further as we proceed. It also implies that this void contained potential quantum energy. Had Moses not been obedient to the Word of God and written as he was inspired by the Holy Spirit, it is highly unlikely that he would have written these words that seemed to be counterintuitive. He wrote thatbefore there was time and space, there was a dark vacuum that contained nothing, but this nothing contained a region of chaos that was constantly moving, and it was contained and had a border - or had a face. Today it makes sense. There couldn't have been any light because there weren't any waves to enable the light to propagate, only quantum fluctuations that were chaotic bursts of continually moving energy - just as Moses wrote! Coincidence?

"...And the Spirit of God was hovering over the face of the waters..."

Here's where things really get interesting. Scripture tells us: *"And the Spirit of God was hovering over the face of the waters."* God provided the mechanism through his Spirit that acted as a catalyst, providing more movement in the form of kinetic energy that changes the vacuum. Theories in quantum electrodynamics postulate that elementary particles interact with ZPEs, resulting in polarization[xxv]. This can be easy to miss if one relies solely on a simple reading of the Scripture without researching the Hebrew meaning. The Hebrew word implies much more than a gentle vibrating motion. It implies that God shook the waters very energetically and caused them to tremble. This is also the first reference to "waters." The scientific connection between the shaking and subsequent creation of heat within a vacuum resulting in the presence of plasma (water) has only recently been established through experiments at CERN[xxvi]. These experiments led to the announcement of the discovery of the long sought after Higgs Boson, commonly referred to as the "God particle" because it was acknowledged that this enabled matter to be turned into mass[xxvii].

The introduction of this shaking of the vacuum containing quantum fluctuations, created an intense amount of kinetic energy. Here we have a classic example as to how God provides the mechanism that brings something out of nothing. In this regard, God Himself is the catalyst. Evolutionists do not have an explanation or any idea of what that the mechanism was or where it came from. Traditionally, they refer to the transition from nothing into something as a Big Bang, based on the expansion of the universe that can be observed. However, as previously stated, there isn't any attempt to explain the matter that their theory needs to be plausible or the mechanism that began the expansion. Their theory requires that a dense piece of matter existed primordially and then something happened to cause it to expand. In the past, they believed that this matter existed with monopoles, but since the search for monopoles led science to believe that monopoles do not exist, plasma physics then provided the explanation. Their explanation was that the original primordial matter was plasma - or a form of pure water.

There are many things about the Big Bang theory that are correct and scientifically supportable - such as the expansion of the universe. But since there are 17 verses in the Old Testament that clearly tell us that God stretched and is continuing to stretch the universe, expansion also supports the creation account. And when scientific data supports two opposing theories, it should be logical to conclude that the data cannot be used to prove either theory.

The waters

The word "waters" has been overlooked by Creationists who seem a bit puzzled by its existence, but it is another astounding scientific assertion. The popular point of view regarding "the waters" is that they were there at the time of creation, since we are not told that God created the waters. Or are we? In a sense, this point of view is nearly as troubling for Creationists as the unexplained matter is for Big Bang theorists. Genesis clearly states that "at the beginning" God created, meaning that before anything existed, God created, and "anything" would include water. As it turns out, again modern day plasma physics holds the explanation regarding the waters being referred to in this verse[xxviii].

In physics, plasma is acknowledged as a liquid and it is often thought of as the fourth state of water. Its distinguishing feature is that its properties are the result of interactions between charged particles - or virtual particles, which can be the result of the transfer of energy from waves to particles. It should be noted here that if the ZPE is in equilibrium, the transfer of energy will not occur. Of course the first condition for system self-organization requires that the nucleus ZPE be driven off thermodynamic equilibrium by abrupt motion. Once again, this is explained by God's hovering. Some have even gone as far as suggesting in their theories that some type of energy flux needs to intersect from a higher dimensional superspace[xxix] in order to drive the nucleus-ZPE off equilibrium.

Experiments in plasma physics have yielded more than interesting results. Heating water through thermal motion well beyond the boiling point where it turns to gas, creates a state where the atomic bonds are broken and result in negatively and positively charged electrons forming a fully ionized gas. At high enough temperatures the gas that was once water becomes completely ionized plasma. Plasma behaves as a formless fluid that is electrically active[xxx]. There are many types of plasmas, and succeeding verses in the Bible talk about God separating the waters from the waters. Suffice it to say at this point, that physics now believes that the entire cosmos (space) exists as a form of plasma - translation: water. Once again, this is another situation where modern science is catching up with Genesis.

If you reverse the process and think of the state of the pre-creation universe, the state of the vacuum would have been the same as the state in a ZPE prior to the existence of plasma. This is not a stretch, since in deterministic physics, all processes can proceed backward as well as forward through time. This approach was utilized in formulating the Big Bang Theory, in which the expansion of the universe was observed and then run backward in time to the point where the expansion began.

In the primeval state there were chaotic bursts of energy. In plasma/ZPE experiments, this state can be created in reverse as done at CERN, by forcing elements to collide within a vacuum. (It should be noted that this work proposes that the collision within the primordial void was achieved through the shaking - or as described in Genesis - the "hovering" over the void by the Spirit of God).

The objective of the collisions at CERN was to break the gluons that keep the quarks together - in other words, blasting the element apart right down to its building blocks. When this happens, the elements no longer exist but the quarks and gluons emerge and form a liquid state, known as quark/gluon plasma. In this high temperature state the plasma behaves as a fluid and not a gas, as one might expect.

The primeval vacuum poses a problem for thermodynamics that states that systems go from an organized state and evolve toward random behavior[xxxi]. However, at the time of the creation, things went the opposite way: from a chaotic or a random state to an organized state. This is not natural and requires interference of some kind, or something over and above the natural - something "supernatural." This presents another problem for Evolutionists who are faced with finding a natural cause - or at least some kind of hypothesis - for not only the singularity but for the mechanism that caused the expansion. This problem is now compounded by the unspoken question of how that mechanism brought order out of chaos. Some spark of hope for these dilemmas of Evolutionists might have occurred in 1977.

In 1977, Ilya Prigogine won the Nobel Prize in Chemistry for defining how systems can go from randomness to coherence - or organization.[xxxii] He discovered that the importation and dissipation of energy into a chemical system could violate the second law of thermodynamics and result in organization. This led to research into something referred to as self-organizing systems. It tries to get around the issue of organization by arguing that real energy was already present and therefore conservation would not be an issue. While conservation of energy is not an issue, either in the Big Bang or Creation model, the organization of energy from chaos to coherence problem lingers on for the Evolutionist but not for the Creationist, who should recognize that the importation of energy into a system could violate the second law of thermodynamics - as Prigogine predicted. In this case, the importation of energy would have occurred when God hovered over the vacuum. And if the Big Bang is relevant, than it must be conceded that at one point organization replaced chaos - not only at the time of the initial expansion but also at the time when Evolutionists believe that gravity pulled matter into clumping (accretion) and formed celestial bodies.

The clumping of matter into celestial bodies is

also in defiance of the Second Law of Thermodynamics; however, Evolutionists employ the force of gravity at this point. But once again, the sudden existence of the force of gravity remains an enigma - even though it was thought to have been understood as one of the four forces in nature. Although Newton's and Einstein's theories about gravity are considered to be the standards, *evolving* physics realizes that there must be a problem with our current understanding, since it is widely believed that gravity seems to be the force that prohibits unification. This is a bit of a conundrum for those who defend Uniformity - especially since gravity resulting in clumps of dust accreting into a body of any kind has not been observed - nor has gravity (without a catalyst) been observed as a mechanism enabling accretion. No matter how you slice it, attributing gravity as the mechanism that formed celestial bodies out of cosmic dust is somewhat of a stretch and cannot be proven by observable science - unless of course, magnetism is applied and properly understood to be the mechanism that drives gravity, as well as the other forces. But then, again, the question arises as to how magnetism appeared out of nowhere.

At the time God was hovering over the vacuum, He also added an immense amount of heat and pressure into the void. Again, He did as Prigogine had argued. He added energy into the system. It is also important to note that verse 2 specifically states that God was hovering over the *face* of the waters. The word "face" makes it clear that God was on the outside of the void. This is important when considering that He added pressure. Had He not been outside of the void, He could not have applied pressure. The addition of pressure is important because physics has discovered that the application of pressure can absorb kinetic energy and convert it into density and pressure waves - both of which are critical for the soundwaves God uses to create light and subsequently, matter.

Pressure waves also provide an environment within a ZPE that can further develop into a show of strong oscillatory patterns called baryonic acoustic oscillations, in other words - sound waves. The development of an environment that accommodates the existence of pressure waves is essential for the eventual production of sound waves and radiation waves (light) - a development that is acknowledged in the next verse. Without the existence of plasma, the propagation and transmittal of soundwaves would not have been possible, since sound cannot travel through vacuums. Pressure waves also are significant because they require that an external force must be present outside of the void in order for the pressure upon the void to transpire. In essence, the necessity of pressure upon the void for the creation of virtual particles *voids* the notion that a natural force within the void is possible for the transition of the quantum fluctuations into waves.

It is truly astounding that every word in verse 2 has significant scientific meaning. Even an advanced physicist, proficient in all the branches of physics, would have a hard time constructing one sentence to describe the entirety of the pre-creation universe along with the applied physics that would eventually bring something out of nothing.

To recap: aside from the topic sentence, the Genesis account tells us what existed primevally regarding the void (vacuum). Then we are also told about the initial actions (perturbations) that God employed when He hovered, which resulted in water (plasma), and then spoke (acoustical waves) and created light, the cosmos, and eventually matter.

Chapter Four

GENESIS 1:3: "Then God said, 'Let there be light'; and there was light."

This verse is also scientifically profound and yet astonishingly simple. It begs the question: what came first? Sound waves or radiation waves? We know that waves transfer energy from one place to another.[xxxiii] Sound waves do this as well as radiation waves. The primary difference between the two is that light is polarized and travels transversely, and in most cases sound is not polarized and travels longitudinally.[xxxiv]

So, if light is traveling transversely and sound is traveling longitudinally, we can postulate that the emergence of the two at the same time was multi-dimensional, as studies in harmonic resonance asserts for waves.

Genesis 1:3 tells us that God spoke light into existence. He spoke into a highly excited field of chaotic energy that He had vibrated, which created a state of kinetic energy and pressure waves that produced the plasma that Moses called water. The plasma is important because without its presence, science will argue that the observable laws of physics do not allow sound waves to propagate through a vacuum - or a void. But physics also tells us that sound waves *do* propagate through plasma. And it is important to understand that sound and light exist in the form of waves that display both similar and different characteristics, and it is also important that the propagation of waves require a source.

In an article published by New Scientist in June of 2004, entitled *Universe started with hiss, not bang* authored by David L. Chandler about the claims of cosmologist Mark Whittle, Chandler states:

> *Cosmologists do not usually think in terms of sound, but this aural picture is a good way to think about the Universe's beginnings, says astronomer Mark Whittle of the University of Virginia in Charlottesville. Whittle has reconstructed the cosmic cacophony from data teased out over the past couple of years from the high-resolution mapping by NASA's WMAP spacecraft of the cosmic microwave background radiation, the afterglow of the hot early Universe.*
>
> *The variations in the cosmic background radiation expose the relative clumpiness of the early cosmos at a variety of different scales. These density variations began as quantum fluctuations in the moments after the big bang, and then propagated out as sonic waves. The denser regions became*

the seeds of galaxies and stars, which is why astronomers are so interested in them.

While at this point, Whittle's theory is not considered to be the standard in Cosmogony, it is a partial confirmation of the Creation Account sequence in Genesis. It errs in placing the Big Bang before the quantum fluctuations and the propagation of sonic (sound) waves but it does recognize the importance of soundwaves in the early universe and acknowledges their existence.

However, Whittle is not the first to recognize that sound waves played an important role in the early universe. NASA also came to this conclusion. Following is NASA's map of the sound wave that propagated through the early universe. Note that NASA also acknowledges that the peaks indicate that harmonics were present. This will become important as we move forward, since harmonics also allow for the resonance of sound waves, which in turn I contend allow for the propagation of infinite multiple dimension. .

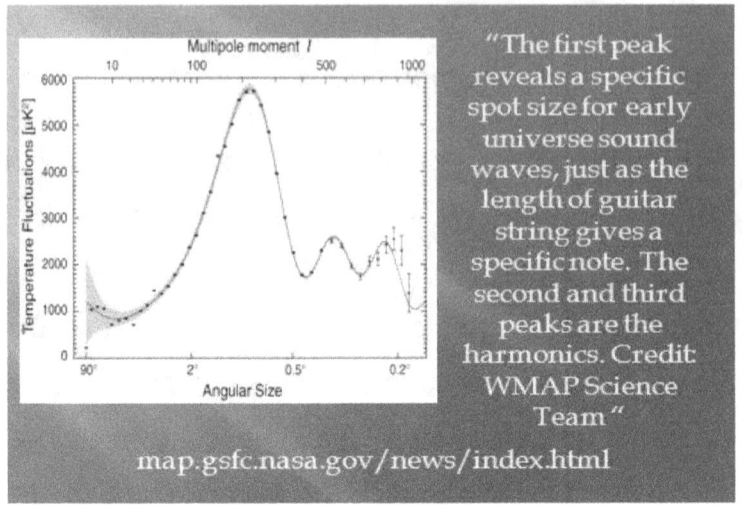

"The first peak reveals a specific spot size for early universe sound waves, just as the length of guitar string gives a specific note. The second and third peaks are the harmonics. Credit WMAP Science Team"

map.gsfc.nasa.gov/news/index.html

In the act of speaking, God sent acoustical waves through the primordial plasma field, and the waves provided the environment that organized some of the energy into light, which developed polarization and would enable the formation of charges and eventually matter. This is the point of transition. It is where a wave propagates through a field of highly excited energy, and in the process, some of the energy is polarized by the current of the wave and particles. This supplies the electrical charges that are necessary for elements (or matter) to come into existence.

Normally sound waves are longitudinal waves that do not exhibit polarization but these waves were highly pressurized and did not exist in an equalized medium, therefore, they would have been both transverse and longitudinal.

Additionally, we also know that at the surfaces of liquids, vibration (or sound waves) can also travel transversely, where the compression is at right angles to the direction of motion. Remember, God hovered over the "face" of the waters, indicating there was a surface to the void. At this point, God had not stretched out the creation and the Big Bang's expansion had not occurred, so there was plenty of surface for the pressure waves to work on the plasma and create incalculable phases of resonance .

Science also tells us that sound waves are generated by a sound source that create vibrations that propagate away from the source in the form of a wave. The Genesis account tells us that there was intense vibration, not only at the time God was hovering but also at the time God spoke. In other words, He (God, the source) created sound waves that propagated through intensely vibrating plasma away from Him (the source), which most likely oscillated the sound waves, since the plasma did not have constant physical properties at this point in time. This type of environment is capable of producing the type of sound wave that emits light and is polarized. Again, this is a problem for Evolutionists that now must be able to explain a source for the primordial sound wave, as well as a source for radiation.

The NASA map previously sited clearly states that the sound waves produced primordially were harmonic in nature. This is also evidence that the waves were capable of traveling transversely - or at different angles. In transverse waves, mixed tones can be produced in the form of harmonics. It should also be noted that the amplitude of the sound also affects the creation of harmonics[xxxv].

Modern experiments have shown that propagating sound waves through a liquid emits light in what is called sonar luminescence or sono luminescence[xxxvi]. As the sound waves travel through an unstable liquid - such as highly ionized plasma - extremely hot bubbles form that burst and emit light. This happens as the waves excite the energy by ultrasonic irradiation, which form multiple bubbles that emit bright flashes of light when they collapse. These bubbles are so hot that they can literally melt steel. Another interesting meaning for the Hebrew word "Yom" that is translated as the word "day" in English, literally means "to be hot" - another strange coincidence where the Genesis account is supported by modern day physics.

It should also be noted that the mechanical qualities of these primordial sound waves created current through the plasma, which gave way to electromagnetism. This characteristic enabled the extant chaos to be organized through polarization, first as light and later as matter.

While Genesis 1:3 is not the creation of matter, it is the first explanation for the creation of turning a virtual particle into a particle that is disbursed throughout the dimensions within a light cone (an interesting phenomenon and worthy of study).

At this point we have a sequence: first there was a void; then God hovered over the void and introduced kinetic energy, through which He brought water/plasma (all of which can be verified through modern experiments); after which He spoke into the plasma, which sent mechanical acoustical (oscillating) waves through the environment capable of producing radiation and electromagnetic waves. These waves propagated through the plasma and created an environment capable of propagating light and sound, which both came into being the first time when He spoke light into existence. God then stretched the light and in so doing, created time and space as well as a means for storing information. The stretching of the light and the phenomena of non-locality, which Einstein referred to as "spooky things at a distance", will be discussed in Part 2.

It is also the point in time when God created

dimension (or space) when He pressurized the waves as He stretched out the creation on Day 2, a crucial point that is overlooked. The stretching that is clearly documented in Genesis, as well as 17 other verses throughout the Bible, is observable today and the basis for the Big Bang Theory that is not discussed in this work. What is not commonly understood is the phenomenon that resulted as a result of the creation being stretched, such as time dilation, harmonic resonance, Doppler shifts, and the non-locality of light.

This newly created environment was the beginning of providing the plasma, the waves and the mechanisms whereby the polarized energy was organized into charges and then into matter, and which also allowed for the recording of information. It is how God brought something out of nothing.

Hence, the scientific search for origins leads directly back to God. If you don't believe in God, you are faced with a scientific dilemma. You must be able to:

- propose a mechanism for adding kinetic energy into a ZPE and creating plasma;
- and explain how this plasma converts into waves without an external source;
- and you have to be able to explain how and where the applied pressure came from to turn chaos into organization without violating the Second Law of Thermodynamics;
- and then explain how did this organization of virtual particles produce light without an external force or without a source to propagate the light;
- and then explain the ability of these waves to create and record information without an intelligent source;
- and then how did the matter we observe today come into existence;
- and how accretion and the four recognized forces of physics (or arguably the one: electromagnetism) came into existence.

But the problems for Evolution do not end

with the above points. Since NASA has documented a sound wave that propagated at the beginning of the universe, Evolutionists must also be able to explain how that sound wave propagated through a ZPE. They must also be able to explain the mechanism for the expansion of the popular Big Bang, or sudden expansion of the singularity, and the organization that developed from the primordial chaos without a defined mechanism or input of external energy.

All of these questions can be answered and unusual phenomena can be understood when looking through the lens of Scripture. Like it or not, the sequence and all the sciences needed to bring the universe into being as we observe it today is detailed in the first few verses of Genesis, we simply did not have the science enable us to understand what God was telling us - at least not until today.

In conclusion, it is very important to understand that when God created, He not only brought forth the universe as we know it; He created the physical laws to accommodate or support His creation. It is why there is such a thing as physical laws that demand that somehow and in some way an irreversible and reliable pattern of operation exists. In a created universe that was intelligently designed, physical laws are explainable. In a random universe that came into existence haphazardly through chaos, it is unlikely that consistent and reliable physical laws would exist, since only randomness and chaos can come from randomness and chaos. In a world where periodic tables are reliable and exquisite fractals and Fibonacci sequence abound, chaos is not what we observe.

Therefore, it is the Evolutionist that bears the burden of answering the many unanswered questions regarding the observable universe and its origins through natural means - unencumbered by supernatural activity. For the Creationist that burden has been lifted.

[i] The familiar title was changed in the 1872 edition from: "On the Origin of Species by Means of Natural Selection, or the Preservation of Favoured (sic) Races in the Struggle for Life." It should be noted that the racist implication of the original title gave credibility to tyrants like Hitler to discriminate against those they deemed to be lower on the evolutionary tree.
http://www.nationalreview.com/article/345274/progressive-racism-paul-rahe

[ii]

https://www.nationalreview.com/nrd/articles/338503/darwins-constitution

[iii] Most believe that Moses wrote the first five books of the Bible, and this work assumes that he did. However, if that is not the case and Moses compiled the literature that had been passed down through generations, the legitimacy of the content is not in dispute or compromised in any way. It is documentable that the first five books have been in existence for nearly four millennia and therefore stand as an argument applicable in this work.

[iv] Hawking, Stephen. *Black Holes and Baby Universes and Other Essays.* Bantam Book 1994. Grand Unification Theories described, pp. 49-53.

[v] Podolny, "Something Called Nothing" (Mir Publ., Mocow, 1986). Richard Morris: "In modern physics, there is not such thing as 'nothing'." (Morris 990:25)
http://www.ldolphin.org/zpe.html.

[vi] https://profmattstrassler.com/articles-and-posts/particle-physics-basics/quantum-fluctuations-and-their-energy/

[vii] "Feynman Diagram" Some Frequently Asked Questions About Virtual Particles math.ucr.edu

[viii]

http://www.pbs.org/wgbh/nova/blogs/physics/2012/10/quantum-foam-virtual-particles-and-other-curiosities/

[ix] https://www.theguardian.com/science/2014/apr/05/einstein-equation-emc2-special-relativity-alok-jha

[x] http://peswiki.com/index.php/Directory:Zero_Point_Energy.

http://www.calphysics.org/zpe.html.

[xi] http://www.newscientist.com/article/dn16095-its-confirmed-matter-is-merely-vacuum-fluctuations.htmlIt's, Article: Confirmed: Matter is merely vacuum fluctuations 20 November 2008 by Stephen Battersby.

[xii] http://dictionary.reference.com/browse/vacuum+fluctuation - vacuum fluctuation - A spontaneous, short-lived fluctuation in the energy level of a vacuum, as described by quantum field theory. Although these variations are violations of the law of conservation of energy, they are tolerated in quantum mechanics by virtue of the uncertainty principle. Such fluctuations are associated with virtual particles.
Morris, Richard 1990. *The Edges of Science.* Prentice Hall Press. Vacuum Physics and Quantum Field Theories, pp. 8, 25, 184; quantum mechanics, pp.19. An investigation into and a report on vacuum physics and quantum mechanics.

[xiii] https://www.YouTube.com/watch?v=C3FCF/HNRyQBizzareDiscovery@CERN

[xiv] http://micro.magnet.fsu.edu/primer/lightandcolor/particleorwavehome.html, Light: Particle or a Wave? by Kenneth R. Spring - Scientific Consultant, Lusby, Maryland, 20657 and Matthew J. Parry-Hill, Robert T. Sutter, and Michael W. Davidson - National High Magnetic Field Laboratory, 1800

East Paul Dirac Dr., The Florida State University, Tallahassee, Florida, 32310.

[xv] http://hyperphysics.phy-astr.gsu.edu/hbase/mod1.html. http://micro.magnet.fsu.edu/primer/lightandcolor/particleorwave.html.

[xvi] http://www.plasma-universe.com/Plasma-Universe.com. http://electric-cosmos.org/indexOLD.htm.

[xvii] http://einstein.stanford.edu/content/relativity/q411.html. http://en.wikipedia.org/wiki/Spacetime.

[xviii] Unless otherwise quoted, The Brown-Driver-Briggs Hebrew and English Lexicon and Strong's Comprehensive Concordance of the Bible are used. Additionally, consultations with Hebrew scholar Pam Elder have been relied on as well and contributed to this work with great insight.

[xix] Stong's Comprehensive Concordance of the Bible #776, "erets", meaning to be from the earth as opposed to heaven or sky. Also see The Brown-Briggs Hebrew and English Lexicon explanation coordinating with Strong's #776, 1b.

[xx] https://en.oxforddictionaries.com/definition/vacuum

[xxi] https://arxiv.org/pdf/quant-ph/9706025.pdf, pg. 1

[xxii] Morris, Richard 1990. *The Edges of Science.* Prentice Hall Press. Vacuum Physics and Quantum Field Theories, pp. 8, 25, 184; quantum mechanics, pp.19. An investigation into and a report on vacuum http://homepages.cae.wisc.edu/~callen/book.html physics and quantum mechanics.

[xxiii] http://aliceinfo.cern.ch/Public/en/Chapter1/newsqm2012.html.
http://en.wikipedia.org/wiki/High_energy_nuclear_physics.

[xxiv] For a interesting comparison of the uniqueness of our planet, Earth, to the rest of the cosmos, see "The Priviledged Planet."

[xxv] J. Reinhardt, B. Muller, W. Greiner, *Quantum Electrodynamics of Strong Fields in Heavy Ion Collisions."* Prog. Part. And Nucl. Phys. 4, 503 (1980).

[xxvi] cds.cern.ch/journal/CERNBulletin/2012/19/News%20Articles/1442988. CERN Bulletin No.18-19/2012-Monday 30 April 2012 "MUCH ADO ABOUT NOTHING - EXPLORING THE VACUUM WITH THE LHC"

[xxvii] https://lasers.llnl.gov/science/understanding-the-universe/plasma-physicshttp://www.reuters.com/article/2012/09/13/science-cern-idUSL5E8KDMQN20120913
http://worldsciencefestival.com/videos/lhc_alice_experiment

http://kdal610.com/news/articles/2012/sep/13/alice-scientists-enter-primeval-plasma-wonderland/
http://en.wikipedia.org/wiki/ALICE_experiment

[xxviii] P.A. Dirac, Roy, Soc. Proc. 126, 360 (1930), Also G. Gamow, *Thirty Years that Shook Physics,* Doubleday, NJ, 1966; cds.CERN.ch (See Printed bulletin #1).

[xxix] S.W. Hawking, *Wormholes in Spacetime* Phys. Rev. D 37(4), 904-910 (1988)

D.H. Freedman, *Maker of Worlds,* Discover 11(7), 46-52 (July 1990) 27

[xxx] Fundamentals of Plasma Physics by James D. Callen, University of Wisconsin, Madison
June 28, 2006 page v

[xxxi] https://lasers.llnl.gov/science/understanding-the-universe/plasma-physics

[xxxii]

http://www.osti.gov/accomplishments/prigogine.html.
http://www.nobelprize.org/nobel_prizes/chemistry/laureates/1977/prigogine-bio.html.

[xxxiii]

http://www.nobelprize.org/nobel_prizes/chemistry/laureates/1977/prigogine-bio.html.
http://en.wikipedia.org/wiki/Wave.

[xxxiv] http://en.wikipedia.org/wiki/Polarization_(waves).
http://www.physicsclassroom.com/class/waves/Lesson-1/Categories-of-Waves.
http://www.physicsclassroom.com/class/waves/Lesson-2/The-Anatomy-of-a-Wave.
http://www2.cose.isu.edu/~hackmart/waves100.PDF.
http://physics.info/light/.

[xxxv] http://www.jhu.edu/signals/listen-new/listen-newindex.htm.
http://en.wikipedia.org/wiki/Simple_harmonic_motion.
http://www.phon.ucl.ac.uk/courses/spsci/acoustics/week1-4.pdf. http://www.dspguide.com/ch11/5.htm.

[xxxvi] http://www.sonoluminescence.com/.
http://www.physics.ucla.edu/Sonoluminescence/sono.pdf. http://techmind.org/sl/.

www.ingramcontent.com/pod-product-compliance
Lightning Source LLC
Chambersburg PA
CBHW050016230526
45470CB00003B/990